Smithsonian

LITTLE EXPLORER

COOL CRICKETS

by Megan Cooley Peterson

PEBBLE
a capstone imprint

Little Explorer is published by Pebble, an imprint of Capstone.
1710 Roe Crest Drive
North Mankato, Minnesota 56003
www.capstonepub.com

Copyright © 2020 by Capstone. All rights reserved. No part of this publication may be reproduced in whole or in part, or stored in a retrieval system, or transmitted in any form or by any means, electronic, mechanical, photocopying, recording, or otherwise, without written permission of the publisher.

The name of the Smithsonian Institution and the sunburst logo are registered trademarks of the Smithsonian Institution. For more information, please visit www.si.edu.

Library of Congress Cataloging-in-Publication Data is available on the Library of Congress website.

ISBN: 978-1-9771-1432-7 (library binding)
ISBN: 978-1-9771-1789-2 (paperback)
ISBN: 978-1-9771-1436-5 (eBook PDF)

Summary: Young readers are introduced to the amazing variety of crickets that live around the world while learning about their behavior, life cycle, and more.

Image Credits
Alamy: Daniel Borzynski, 13, Michal Fuglevic, 7; Minden Pictures: Alex Hyde, 23, Emanuele Biggi, 29, Manabu Tsutsui, 25; Newscom: Anthony Bannister/NHPA/Photoshot, 19, David Element/NHPA/Photoshot, 11; Science Source: Scott Camazine, 21, Ted Kinsman, 9; Shutterstock: Alexander Sviridov, 1, 13 (middle left), chinahbzyg, 13 (middle right), F_studio, 4, Jacob Hamblin, 7 (middle), Marek Velechovsky, 8, Melinda Fawver, 5, 7 (bottom), Muhammad Naaim, cover, prapann, 7 (top), Rudmer Zwerver, 27, SIMON SHIM, 15, tea maeklong, 2, Unknown Photographer, 13 (top); Wikimedia: flickr/Judy Gallagher, 16

Editorial Credits
Editor: Abby Huff; Designer: Kyle Grenz; Media Researcher: Tracy Cummins; Production Specialist: Katy LaVigne

Our very special thanks to Gary Hevel, Public Information Officer (Emeritus), Entomology Department, at the Smithsonian National Museum of Natural History. Capstone would also like to thank Kealy Gordon, Product Development Manager, and the following at Smithsonian Enterprises: Ellen Nanney, Licensing Manager; Brigid Ferraro, Vice President, Education and Consumer Products; and Carol LeBlanc, Senior Vice President, Education and Consumer Products.

All internet sites appearing in back matter were available and accurate when this book was sent to press.

Printed and bound in the United States of America.
PA99

Table of Contents

Chirping Crickets..................................... 4
Field Crickets.. 6
Ground Crickets..................................... 10
Sword-Tail Crickets 12
Bush Crickets....................................... 14
Tree Crickets 18
Scaly Crickets..................................... 22
Ant Crickets....................................... 24
Mole Crickets..................................... 26
Cave Crickets..................................... 28

Glossary 30
Critical Thinking Questions....................... 31
Read More... 31
Internet Sites..................................... 31
Index.. 32

Words in **bold** are in the glossary.

Chirping Crickets

Crickets are the rock stars of the **insect** world. Males rub their wings together to "sing." Their chirps help attract females. They also sing to warn other male crickets to stay away.

About 2,400 kinds of crickets live on Earth. Crickets help the Earth in many ways. They eat dead plants. They add **nutrients** to the soil. This helps plants grow. Crickets are also food for other animals.

A bell cricket raises its wings and rubs them together to chirp.

A Cricket's Body

A cricket has three body parts and six legs. Its **antennae** are longer than its body. Two big eyes let it see in many directions at once. A cricket's ears are in its front legs. Strong back legs make the cricket a great jumper.

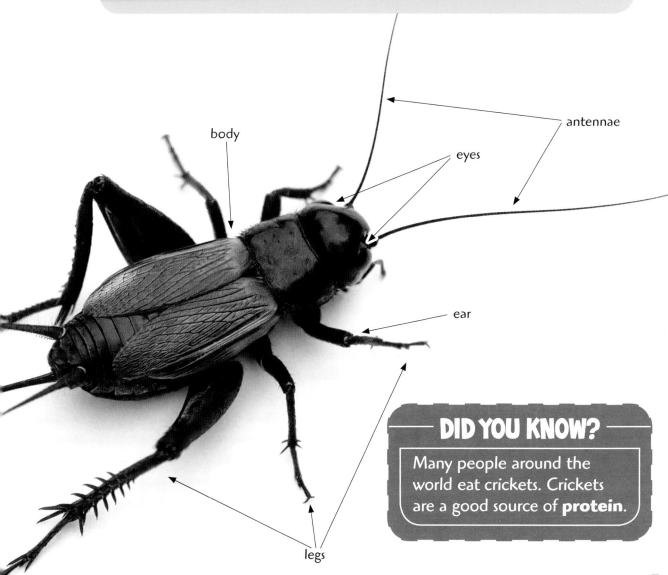

body, eyes, antennae, ear, legs

DID YOU KNOW?

Many people around the world eat crickets. Crickets are a good source of **protein**.

Field Crickets

Number of species: 500
Found: Worldwide except Antarctica
Length: 0.5 to 1.5 inches (1.3 to 3.8 centimeters)

Take a walk through a grassy area. You might spot a field cricket. These crickets are black or brown. Almost all field crickets live on the ground. Some live under logs, leaves, and rocks. Some even live inside buildings and in piles of garbage!

The Japanese **burrowing** cricket lives underground. It moves dirt one mouthful at a time to dig a burrow. The dirt home keeps it safe from snakes, frogs, lizards, and other insects.

European field cricket

Crickets, Grasshoppers, and Katydids

Crickets are related to grasshoppers and katydids. These insects have long back legs for jumping. They all make sounds. But there are differences. Crickets and katydids have longer antennae than grasshoppers. Most crickets can't fly. Grasshoppers can. Grasshoppers and katydids make buzzy sounds. Cricket chirps are more musical.

cricket

grasshopper

katydid

DID YOU KNOW?
Crickets were named for their chirps. People thought their songs sounded like the word *cricket*.

DID YOU KNOW?
Field crickets eat their body weight in food each day.

Field crickets are not picky eaters. They eat both insects and plants, and they also eat dead bugs. Field crickets that come into houses will feed on clothing, rubber, and paper.

When many field crickets live in one place, they can cause problems. They may eat and damage young plants. But crickets are not usually a problem for farms or gardens.

How Do Crickets Sing?

Male crickets have a scraper on the bottom of one front wing. The other front wing has 50 to 250 teethlike bumps. The cricket rubs the scraper along the teeth. It makes a chirping sound. Cricket chirps don't all sound alike. Each **species** of cricket has its own song.

A close-up view of the teethlike bumps on a cricket's front wing

Ground Crickets

Number of species: 300
Found: Worldwide except Antarctica
Length: Up to 0.5 inch (13 cm)

Is that a baby cricket jumping around? No, it's a ground cricket! They look like tiny field crickets. Most ground crickets have dark bodies. They blend in with the dirt. **Predators** don't see them.

Ground crickets live near water in wet, grassy areas. Many live in rain forests. Most crickets are active at night. But some ground crickets come out during the day.

DID YOU KNOW?
Some kinds of Australian ground crickets don't have ears! The males are wingless. They can't chirp. So the crickets don't need ears.

Sword-Tail Crickets

Number of species: 490
Found: Worldwide except Antarctica
Length: 0.15 to 0.28 inch (0.4 to 0.7 cm)

Female crickets have a needlelike part at the end of their bodies. It lays eggs. The female sword-tail cricket's end part is shorter and curved. People say it looks like a sword. Sword-tails live on plants and near water.

It's hard to miss the handsome trig. This sword-tail cricket is colorful. Its body is red and black. Its legs are cream colored. The male sings through the day and night.

DID YOU KNOW?
Female crickets stab their end part into plants, bark, or dirt. Then they lay their eggs inside one at a time.

A Cricket's Life

Female crickets lay eggs. The eggs hatch after about two weeks. Young crickets are called **nymphs**. They look like tiny adult crickets. Nymphs **molt** many times as they grow. Adult crickets live about a year.

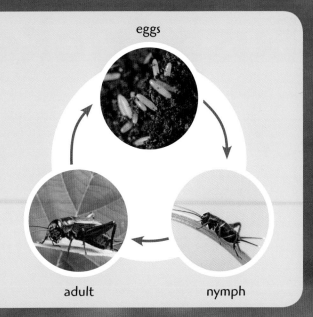

eggs / adult / nymph

The female handsome trig has a short, curved part at the end of her body that lays eggs.

Bush Crickets

Number of species: 500
Found: Worldwide, mostly in tropical areas
Length: 0.35 to 1.4 inches (0.9 to 3.5 cm)

Bush crickets are at home anywhere that's leafy. They live in bushes. Some crawl on herbs and in trees. These insects live in wet, tropical places. They eat leaves, flowers, and fruit.

Like all crickets, bush crickets bite and chew with **mandibles**. They work like teeth. Bush crickets also have tiny armlike parts near the mandibles. They use these parts to hold their food.

common bush cricket

DID YOU KNOW?

In the United Kingdom, the name *bush cricket* is used for katydids.

Some male restless bush crickets have an odd trick during **mating**. They let the females eat their wings! This snack keeps the female busy. Or else she might leave and find a different male. But don't worry about the males. Restless bush crickets don't sing. They don't need their wings.

DID YOU KNOW?

Crickets are seen as signs of good luck in many cultures. In China, some people keep them as pets.

The wings on this male restless bush cricket have been chewed on by a female.

Tree Crickets

Number of species: 172
Found: Worldwide except Antarctica
Length: 0.39 to 0.87 inch (1 to 2.2 cm)

Tree crickets look like walking leaves. They are pale green or white. Males have wings shaped like paddles. Tree crickets live in trees and shrubs. They blend in with their homes.

Some male tree crickets have a trick to make their songs louder. First the male chews a hole in a leaf. He sticks his head through it. His wings cover the hole as he trills. The sound gets louder as it travels through the hole.

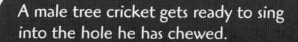
A male tree cricket gets ready to sing into the hole he has chewed.

DID YOU KNOW?
Tree crickets feed on tiny bugs called **aphids**. Aphids can damage plants and crops.

The snowy tree cricket lives in North America. Its chirps can be used to tell how warm or cold the air is. One way is to count the number of chirps in 15 seconds. Then add 40. The new number is close to the temperature in Fahrenheit. Snowy tree crickets chirp faster in warm weather. They chirp slower in cool weather.

Going Silent

On the Hawaiian island of Kauai, most male crickets have gone silent. One kind of fly came to the island in the 1990s. The flies dropped **maggots** onto singing male crickets. The maggots killed the crickets. By 2003, most males were no longer able to sing. Being silent keeps them alive.

DID YOU KNOW?

All cricket songs change with the temperature. The snowy tree cricket's chirps are easier for people to count.

Scaly Crickets

Number of species: About 375
Found: Worldwide except Antarctica
Length: 0.2 to 0.5 inch (0.5 to 1.3 cm)

Scaly crickets look different from most crickets. These tiny jumpers have powdery scales. The scales easily rub off. This might help them get away from predators. Male scaly crickets only have short front wings. Females don't have any. Many scaly crickets live near water in tropical places.

Slosson's scaly cricket has a buzzy chirp. Each chirp has up to 25 pulses. But the chirp lasts less than half a second!

A scaly cricket from the United Kingdom

DID YOU KNOW?

The scaly cricket is one of the United Kingdom's rarest insects. These crickets live on stony beaches. Most people never see them in the wild.

Ant Crickets

Number of species: About 70
Found: Australia, Europe, North America, and Southeast Asia
Length: 0.06 to 0.20 inch (0.15 to 0.5 cm)

Ant crickets move into ant nests and don't move out. Why do they live with ants? For the food! An ant's body makes oils. Ant crickets eat those oils. These tiny, wingless crickets will eat the oil right off the ants. They also steal food that ants bring into the nest. Ant crickets aren't picky. They will live with many kinds of ants.

An ant cricket walks among adult ants and ant larvae.

DID YOU KNOW?

Ants will often attack when an ant cricket chews on them. The crickets quickly run away, but sometimes they get eaten.

Mole Crickets

Number of species: About 100
Found: Worldwide except Antarctica
Length: 0.75 to 2 inches (1.9 to 5 cm)

Mole crickets are digging machines. Their short front legs have claws. They easily push dirt aside. Mole crickets spend most of their lives underground. They live in burrows and dig tunnels. They can dig tunnels more than 20 feet (6 meters) long each night. Mole crickets eat plants, worms, and small bugs.

DID YOU KNOW?

Mole crickets can cause problems for people. They damage lawns and kill vegetable seedlings.

Cave Crickets

Number of species: 250
Found: Worldwide except Antarctica
Length: Up to 2 inches (5 cm)

Is that a spider creeping on the cave wall? Or is it a cave cricket? These crickets have very long legs. Their backs are round. People sometimes mistake them for spiders. Cave crickets don't have wings. But they are skilled jumpers. Some can leap many feet into the air.

DID YOU KNOW?

Cave crickets love oatmeal! To study these crickets, scientists set out piles of oatmeal. Cave crickets follow the smell.

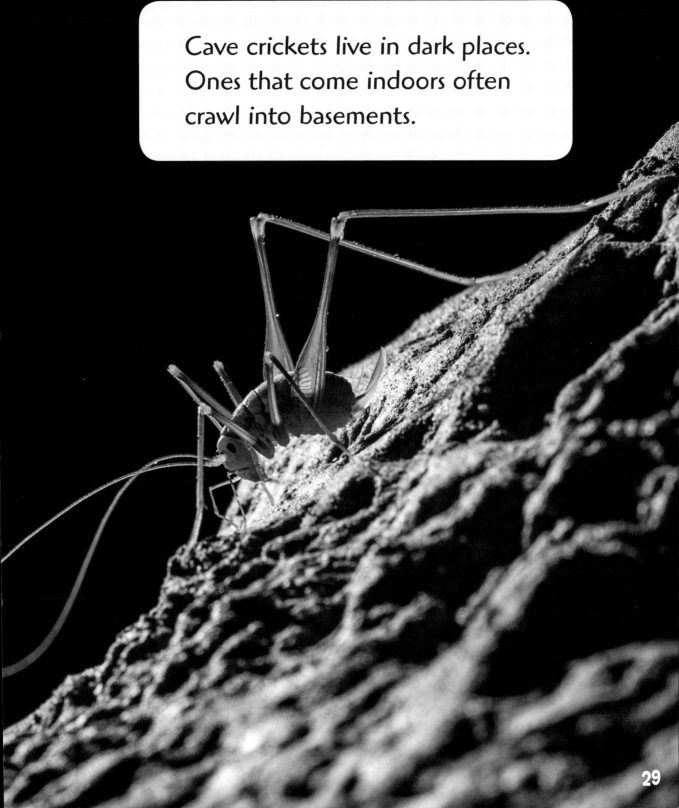

Cave crickets live in dark places. Ones that come indoors often crawl into basements.

Glossary

antenna (an-TEH-nuh)—a feeler on an insect's head used to smell and sense movement

aphid (AY-fid)—an insect that sucks plant juices

burrow (BUHR-oh)—a hole in the ground that an animal makes

insect (IN-sekt)—a small animal with a hard outer shell, six legs, three body sections, and two antennae

maggot (MAG-uht)—a young stage of flies in between egg and adult

mandibles (MAN-duh-buhlz)—strong mouthparts used to chew

mate (MATE)—to join together to produce young

molt (MOLT)—to shed an outer layer of skin

nutrient (NOO-tree-uhnt)—a substance needed by a living thing to stay healthy

nymph (NIMF)—a young form of an insect; nymphs change into adults by shedding their skin many times

predator (PRED-uh-tur)—an animal that hunts other animals for food

protein (PROH-teen)—a part of food that builds strong bones and muscles

species (SPEE-seez)—a group of living things that can reproduce with one another

Critical Thinking Questions

1. Why do you think each species of cricket has its own song? What would happen if all crickets made the same chirping sound?

2. Crickets are closely related to grasshoppers and katydids. How are they alike? How are they different?

3. Why are field crickets black and brown? Why are tree crickets green? How do the different body colors help each cricket?

Read More

Gish, Ashley. *Crickets*. Mankato, MN: Creative Education, 2018.

Martins, Dino. *You Can Be an Entomologist!* Washington, D.C.: National Geographic, 2019.

Perish, Patrick. *Crickets*. Minneapolis: Bellwether Media, 2019.

Internet Sites

Bug Facts: Crickets
https://www.bugfacts.net/cricket.php

DK Find Out!: Grasshoppers and Crickets
https://www.dkfindout.com/us/animals-and-nature/insects/grasshoppers-and-crickets/

Songs of Insects: Old Field Singers
Listen to the songs of bugs often found in grassy fields, including crickets.
http://songsofinsects.com/insect-song-interactive

Index

adaptations, 20
antennae, 5, 7
ants, 24, 25
aphids, 19

bodies, 5, 12
burrows, 6, 26

chirping, 4, 7, 9, 10, 17, 18, 20, 21, 22
colors, 6, 10, 18

digging, 26
dirt, 6, 10, 12, 26

ears, 5, 10
eggs, 12, 13
eyes, 5

flying, 7
food, 4, 8, 9, 14, 19, 21, 26

grasshoppers, 7

jumping, 5, 28

katydids, 7, 15

leaves, 6, 14, 18
legs, 5, 7, 12, 26, 28
life cycles, 13

maggots, 20
mandibles, 14
mating, 17
molting, 13

nutrients, 4
nymphs, 13

plants, 4, 9, 12, 26
predators, 6, 10, 22, 25
protein, 5

scales, 22

temperature, 20, 21

water, 10, 12, 22
wings, 1, 9, 10, 17, 18, 22, 24, 28